THE BEST BASIC REFRIGERATION OPERATOR EXAM PREP COURSE

(Boiler Plant Series Book 2)

by Dan Ringo

FOREWORD

Welcome to the 2nd book in the Best Boiler Plant Series. This edition takes a view at Basic Refrigeration Operation in the successfully proven format used to pass license examinations in various municipalities.

Refrigeration and the removal of heat is simply the flipside or opposite of the application of heat. Many of the terms discussed in Best Boiler Operator Exam Prep will be reinforced here for your benefit.

The field and study of refrigeration will continue to grow as well as the job opportunities as more licensed operators leave the market due to retirement. Skilled trades are currently getting a lot of attention from politicians and media outlets and hence are becoming more attractive to those individuals looking for an alternative career than traditional college.

My career started in skilled trades and the first occupational license I ever attained was a City of Detroit 3rd Class Refrigeration Operator. From that license I was able to further my studies in refrigeration to eventually attain a 1st Class Refrigeration License, a Chief Powerhouse License, an Unlimited Refrigeration Journeyman and a Mechanical Contractor License. Through all my twenty-five years experience I have found that a basic and solid foundation is essential to passing an exam and furthering my understanding of the material.

The format presented here is the exact same format I have used myself to acquire my many licenses as well as help countless other students do the same. I am excited to bring you this material and hope you use it as intended to further your career goals.

Finally, I dedicate this book to my father Charles Vernon Ringo Sr, who inspired and encouraged me to give this trade a try almost thirty years ago. Dad, I know you are proud!

TABLE OF

CONTENTS

INTRODUCTION

This book will deliver the complete basic understanding needed for an entry level refrigeration license examination. The information following is presented in a conversational manner with unorthodox examples and phrases to simplify the topics and content. It still includes mandatory terminology and codes related to the field of refrigeration.

As you study refrigeration keep in mind that every component's function is listed or described in its name. For example an evaporator is called an evaporator because that is where the refrigerant evaporates or turns from a liquid to a vapor. A condenser is where the refrigerant condenses or changes from a vapor back into a liquid. The suction line is the line that compressor draws or sucks the refrigerant vapor from the evaporator and so on and so on.

When taking an exam or asked a question by an examiner about a component think first about the name and hence its function.

In your approach to the material and studying for your exam the student should prepare to discuss all of the following chapters by answering these questions about each item.

1

The Best Basic Refrigeration Operator

1. What is it?

2. Where is it located?

3. How does it work?

4. What is its purpose?

5. When is it used?

6. What are some common problems it can cause?

7. What are the remedies to some of the common problems it can cause?

8. What are the types?

Here is an example: Evaporator.

Q. What is an evaporator?

Ans. An evaporator is the cooling coil of the refrigeration system.

Q. Where is the evaporator located?

Ans. It is located between the system's metering device and the compressor on the low-side/low-pressure side of the system.

Q. How does an evaporator work?

Ans. An evaporator absorbs heat from the space or product to be cooled and transfers that heat to the liquid refrigerant traveling through the inside of its tubes. As the liquid refrigerant travels through the evaporator tubing it picks up the heat and begins to boil. It is the boiling of the refrigerant that absorbs the greatest heat from the conditioned space and product and hence provides the greatest cooling. An evaporator must be designed in a manner that allows for the most efficient heat transfer from the space or product through to the liquid refrigerant inside of it.

Using the What, Where, When, Why, How, Problems, Types and Remedies format to learning this material will surely have you prepared to discuss the examination information in any manner presented.

Apply this for all subject matter in this manual and you will be ready to pass your examination. You will see many students studying for an exam are looking for a magic bullet that will teach them by reverse osmosis. You may be thinking well does not this book purport to do the same. Absolutely not! This book does boil down the essentials but also attempts to reshape your thinking and approach to the subject matter regarding basic refrigeration operation and raise a student's confidence in discussing the subject matter with anyone especially an examiner.

So with that being said, let's get into refrigeration, its components, and principles! Are you ready?

HEAT AND HEAT TRANSFER

Stop! Don't go any further. I need your attention. You are looking to learn about what to know in passing a basic Refrigeration Operator License Exam. Well I need you to fully understand the two laws that govern both Boilers as well as Refrigeration. The first is the 1st Law of Thermodynamics.

The 1st Law says, "*Energy cannot be created or destroyed but just transferred from one form to another*."

The 2nd Law says, "*Temperature always downgrade flowing from a hotter substance to a colder one.*"

It is imperative to grasp the definition of heat and energy. Heat is simply a form of energy. Chemical, electrical and mechanical are all forms of energy. Now energy is the ability to do work. You need energy to do work. You need energy to get work done. Now, when a substance absorbs heat it is absorbing energy and therefore the substance is positioned to do work. And the work is cooling off a space, product or part of a process.

Refrigeration is simply the method of removing heat. The basic principle of refrigeration is the transfer of heat from a warm substance to a colder substance. That is in accordance with the Second Law of Thermodynamics. The 2nd Law states simply that heat downgrades. Heat flows from a hotter substance to a colder substance. So, objects that are chilled do not receive "cold" but they lose heat (sensible and latent) until they reach the desired temperature.

<u>Example</u>: In other words, instead of putting cold into a refrigerator for cooling, heat is taken out. The more heat removed, the colder the inside of the refrigerator becomes. As the heat is removed, the temperature drops. Low temperature indicates low heat content. Cold is nothing more than the absence of heat, and refrigeration is just simply a method of removing heat.

In understanding refrigeration is simply the removal of heat it is important to understand that heat is a form of energy. Every object contains heat and heat energy. If all heat is removed from an object, the temperature of the object falls to absolute zero. This is theoretically -459F. Every object contains heat energy in both quantity and intensity.

Refrigeration is the process of cooling a space, substance, or system to lower and/or maintain its temperature below the ambient one. In other words, refrigeration means artificial cooling.

Q. What is heat and heat transfer?

Ans. Heat is a form of energy. There are other forms of energy such as light, mechanical, electrical and chemical. Energy is the ability to do work. For anything or anyone to perform tasks either physically or mentally they must first possess the energy to do so. Energy in a state or form in addition to being chemical or electrical is either at rest (potential) or active (kinetic). When energy is doing work it is said to be kinetic or in motion. When an object is at rest or not moving it is said to have potential. It is the manipulation or transferring from potential to kinetic that gets work done.

Q. What are the forms of heat?

Ans. *Sensible* and *Latent*

Sensible: the temperature that can be measured on or with a thermometer.

Super(heat): the temperature that is above the saturation point. It is the heat that added above the saturation point of a substance. Superheat can be measured with a thermometer. If heat is added after a liquid changes into a vapor it is called superheat. And remember heat is a form of energy. So if more heat is added to the substance that substance now is super.

Sub-cooling(heat): the temperature that is below the saturation point. It is the heat removed below the saturation point of a substance. This heat removal can be measured on or with a thermometer.

Latent Heat:Latent heat is the amount of heat needed to change the state of a substance.

Q. What are the forms of Latent?

Ans. *Evaporation, Sublimation, Condensation, Fusion*

Q. What is evaporation?

Ans. Amount of heat needed to change the state of a liquid to a vapor.

Q. What is sublimation?

Ans. Amount of heat needed to go from a solid to a gas.

Q. What is condensation?

Ans. Amount of heat removed to change a substance from a vapor to a liquid.

Q. What is fusion?

Ans. Amount of heat needed to change a liquid to a solid.

*Q. What is saturation?

Ans. Saturation is when a substance's temperature is also at its corresponding pressure.

Heat Transfer

Now that we have that out of the way it is equally important to comprehend the three methods of heat transfer; *conduction, convection* and *radiation*.

Conduction is a form of heat transfer that occurs from direct contact of the medium giving off heat and the medium receiving heat. The two objects are physically in contact with each other. The best example of this would be skin to skin contact with another person who is at a different temperature. Immediately your body either gives off heat or feels warmer because it is absorbing heat from the warmer body.

Convection is a form of heat transfer that occurs through a flow of air. The air is either cooled or warmed and that air either removes or supply heat or cold to a substance. The heat transfer is indirect but because of the potential velocity of the flow of fluid it can be a much faster mode of heat transfer. The best example of this would be a convection stove. The use of a fan sends heat waves to transfer heat to the items to be cooked.

Radiation is a form of heat transfer that occurs through being in direct sight of the medium either heating or removing heat from another medium or body. The best example of this would be the sun itself. To feel the heat of the sun you must stand in direct line of sight of it.

Heat measurement

British Thermal Unit or BTUs:

Q. What is a BTU?

Ans: The British thermal unit (Btu or BTU) is a traditional unit of heat; it is defined as the amount of heat required to raise the temperature of one pound of water by one-degree Fahrenheit. It is also part of the United States customary units.

Q. How many BTUs are in a ton of refrigeration?

Ans. 12,000 BTU (British Thermal Units) is equal to 1 ton. For heating, BTUs are the amount of thermal energy added to an area. For cooling, BTUs are the amount of thermal energy removed from an area.

Q. Why is refrigeration measured in tons?

Ans. The unit of measurement refers to the refrigeration system's capacity. It's measured in tons to reflect the amount of heat your refrigeration system can remove from a space in an hour.

Q. What is entropy?

Ans. The entropy of an object is a measure of the amount of energy which is unavailable to do work.

Q. What is enthalpy?

Ans. When a substance changes at constant pressure, enthalpy tells how much heat and work was added or removed from the substance. Enthalpy is similar to energy, but not the same. When a substance grows or shrinks, energy is used up or released.

REFRIGERATION

Now we are ready to get into what is refrigeration? Remember, refrigeration is the removal of heat from a place from where it is unwanted to a place where it is less objectionable like outdoors. Refrigeration is the business of moving heat. The evaporator brings heat into the system. The compressor moves that heat. The condenser rejects the heat so that the cycle (refrigeration) can repeat itself all over again and again.

Q. What is refrigeration?

Ans. Refrigeration is simply the method of removing heat. The basic principle of refrigeration is the transfer of heat from a place where heat is unwanted to a place that does not matter; usually outside or outdoors.

Q. Where do you find refrigeration?

Ans. You find refrigeration everywhere. The applications where refrigeration is used are residential, commercial and industrial in small and large applications. You can find refrigeration in grocery stores and in college chiller plant installations. Anywhere cooling is needed and it needs to be controlled you will find a mechanical refrigeration process.

THE BEST BASIC REFRIGERATION OPERATOR

Q. How does mechanical refrigeration work?

Ans. The compressor creates a pressure differential which allows fluid to flow from the high side to the low side. As this flow is created liquid refrigerant enters the evaporator which is located directly in sight of or in proximity of the space or product you want cooled. The refrigerant that is a much lower temperature and pressure than the items you want cool to absorb this heat. As the refrigerant absorbs the heat it temperature rises until it boils and changes from a liquid into a vapor. It is the changing of liquid into a vapor that does the most heat absorption or cooling effect. The compressor which acts like a pump creating a pressure difference sucks the refrigerant vapor from the evaporator and compresses it thus raising its temperature and pressure. Once the compressor does that it sends the hot refrigerant vapor to the condenser where it gives off its heat to the cooling medium (air or water) and turns back into a liquid. Once it is turned back into a liquid it is ready to be used again in the evaporator to begin the cooling process all over. This process repeats until the conditioned space or product reaches its desired temperature and the compressor shuts off until it needs to run again.

Think about it in terms that the evaporator brings in heat. The compressor moves the heat. The condenser rejects the heat out of the system. Brings, Moves, Rejects!

Q. What is the purpose of refrigeration?

Ans. The purpose of refrigeration is solely dependent upon the application it is used. In refrigeration applications it is to provide comfort if the desired intent is to cool space for human occupation. In industry it can be to preserve food. Refrigeration can even be used in refrigeration systems to remove non-condensable gases that accumulate in the system and need to be purged out. However, the purpose of refrigeration is to absorb heat from one place and remove to where it is not a problem. Remember the 1st Law of Thermodynamics? The 1st Law says, "Energy cannot be created or destroyed but transferred from one form to another." And since heat is a form of energy, the heat absorbed in the evaporator can only be transferred to some other place. It is not destroyed. It is simply removed and put somewhere else where it dissipates. All of the heat absorbed in the evaporator must be removed through the condenser for the refrigeration system to operate efficiently. *NOTE* The condenser is sized slightly larger than the evaporator because it removes or rejects all of the heat that the evaporator brings in but also the heat of compression from the compressor.

Now there are tons of equations and formulas about converting Fahrenheit to Celsius and Celsius to Fahrenheit. But you will not be required to discuss temperature scale conversions so I will not waste your time here.

Q. When is refrigeration used?

Ans. Refrigeration is used when close control of a temperature relative to a space, product or process is desired if not needed. Time is also a major factor used to determine whether refrigeration is to be used. Mechanical refrigeration speeds up the heat transfer necessary to reach a desired temperature.

- Ex: You own an ice cream manufacturing process. Your company ships ice cream all over the country. After you make the flavor you need to immediately chill it so it can harden. You need to get the mixture below 20F. And you need to ship the ice cream in cold storage vehicles, so it does not melt and affect its quality.

Refrigeration in the sense discussed here describes Mechanical Refrigeration. Remember the definition of refrigeration? It is the removal of heat from a place where it is unwanted to a place where it is unobjectionable. You do not need all of the components of a refrigeration system to do that. A cool breeze from wind will remove heat from your body and move it to further downstream where you have no concern. In the basic understanding that is, in fact, refrigeration. The cooler wind blew across your body. Your body being at a higher temperature tried to warm up that breeze and hence drop in temperature. You feel that drop in temperature which sends chills through your body.

However, to speed up that cooling or to cool more people from that breeze we need more air flow and a cooler wind to produce the desired temperature. To do this we need the traditional components of mechanical refrigeration to accomplish this.

Q. What are some of the problems with refrigeration?

Ans. Mechanical refrigeration problems stem from the operability of the equipment itself. There is quite a bit of maintenance required to ensure that the systems operate in the safest and most efficient manner possible and in accordance with regulations. Mechanical refrigeration system operations can be costly from an electrical usage or kilowatt price.

Q. What are remedies or preventative measures for the problems refrigeration cause?

Ans. The innate problem of cost that is built-in with mechanical refrigeration can be best offset with a solid preventative maintenance and sustainability program. Making sure operators perform scheduled maintenance per the manufacturer's recommendations is essential. Sustainability efforts to keep energy use costs in line with the owner's budget goes a long way to assist in that arena.

Q. What are some types of refrigeration systems?

Ans. Two Stage and Cascade

Two Stage:

Two Stage systems refer to a mechanical refrigeration system with two compressors. The first compressor is responsible for taking or sucking the vapor from the evaporator and compressing it slightly raising the refrigerant vapor's temperature and pressure. However, the discharge of the first stage compressor serves as the suction for the booster or 2nd stage compressor. The booster or 2nd Stage compressor raises the refrigerant vapor's pressure and temperature above the condensing temperature so that the refrigerant can turn back into a liquid after exhausting its latent heat of condensation. Two stage systems are used with refrigerants that have an extremely low boiling temperature and one stage compression will not raise its temperature and pressure above the condensing medium used.

Cascade System:

A Cascade System has a significantly large temperature and pressure difference, one vapor compression refrigeration cycles become impractical. One of the solutions for such cases is to perform cooling in two or more stages (i.e., two or more cycles), who work in the series. These refrigeration cycles are called cascade refrigeration cycles. Thus, cascade systems used in order to obtain high temperature differences between the heat source and heat sink and are applied at temperatures from -70C -100C. Application of the three-stage compression system for boiling temperatures below -70C is limited, due to difficulties with the refrigerant temperature of freezing. The irrelevance of a three-stage vapor compression systems can be avoided by applying a cascading steam-compression refrigeration machine.

In these systems with several evaporators can be used in any stage of compression. Refrigerants used at each stage may be different and selected for optimum performance at a given temperature of the evaporator and condenser. Ordinary one pump, mechanical cooling system condensing units can reach temperatures of the order. At lower temperatures required then cascade refrigeration systems must be used. A two-stage cascade system uses two refrigeration systems, connected in series to achieve a temperature of about -85C. There are single compressor systems, which can reach temperatures lower than, but they are not widely used.

These systems are sometimes called auto cascade systems. The main disadvantage of these systems is that it requires the use of a proprietary blend refrigerant.

This characteristic results in three of problems connected with: A leak in the system can easily lead to the loss of only some of the refrigerant components of the mixture of the refrigerant mixture, which consists of different types of refrigerants with different boiling points), as a result of an imbalance in the ratio of the remaining refrigerants. To return the system to function, all the other refrigerant should be replaced with a new and potentially expensive cost, to ensure the correct combination of attitude.

The mixture is proprietary and may not be readily available from traditional refrigerant power sources and therefore may be difficult and expensive. These types of cascade systems are not widely used, it is difficult to find highly skilled field service staff who are familiar with repair and maintenance procedures. Of course, these and other issues may cause unwanted costs and downtime.

Q. What are the gas laws pertaining to refrigeration?

Ans. The physical laws that describe the properties of gases, including Boyle, Gay-Lussac and Charles' laws.

Boyle's Law: The Pressure-Volume Law

Boyle's law or the pressure-volume law states that the volume of a given amount of gas held at constant temperature varies inversely with the applied pressure when the temperature and mass are constant.

Charles' Law: The Temperature-Volume Law

This law states that the volume of a given amount of gas held at constant pressure is directly proportional to the Kelvin temperature. Charles' law describes the relationship between temperature and volume at a constant pressure.

Gay-Lussac's Law: The Pressure-Temperature Law

This law states that the pressure of a given amount of gas held at constant volume is directly proportional to the Kelvin temperature.

One of the most amazing things about gases is that, despite the wide differences in chemical properties, all the gases more or less obey the gas laws. The gas laws deal with how gases behave with respect to pressure, volume, temperature, and amount.

Pressure

Gases are the only state of matter that can be compressed very tightly or expanded to fill a very large space. <u>Pressure</u> is force per unit area, calculated by dividing the force by the area on which the force acts. The earth's gravity acts on air molecules to create a force, that of the air pushing on the earth. This is called <u>atmospheric pressure</u>.

The units of pressure that are used are <u>pascal</u> (Pa), standard atmosphere (atm), and torr. 1 atm is the average pressure at sea level. It is normally used as a standard unit of pressure. The SI unit though, is the pascal. 101,325 pascals equals 1 atm.

Refrigeration Conclusion

Refrigeration in the sense discussed here describes Mechanical Refrigeration. Remember the definition of refrigeration? It is the removal of heat from a place where it is unwanted to a place where it is unobjectionable.

Refrigeration is simply the method of removing heat. The basic principle of refrigeration is the transfer of heat from a warm substance to a colder substance. That is in accordance with the Second Law of Thermodynamics. The 2nd Law states simply that heat downgrades. Heat flows from a hotter substance to a colder substance. So objects that are chilled do not receive "cold" but they lose heat (sensible and latent) until they reach the desired temperature.

You find refrigeration everywhere. The applications where refrigeration is used are residential, commercial and industrial in small and large applications. You can find refrigeration in grocery stores and in college chiller plant installations. Anywhere cooling is needed, and it needs to be controlled you will find a mechanical refrigeration process.

The purpose of refrigeration is solely dependent upon the application it is used. In refrigeration applications it is to provide comfort if the desired intent is to cool space for human occupation. In industry it can be to preserve food.

"Refrigeration principles" can even be used in refrigeration systems to remove non-condensable gases that accumulate in the system and need to be purged out. However, the purpose of refrigeration is to absorb heat from one place and remove to where it is not a problem. Remember the 1st Law of Thermodynamics? The 1st Law says, "Energy cannot be created or destroyed but transferred from one form to another." And since heat is a form of energy, the heat absorbed in the evaporator can only be transferred to some other place. It is not destroyed. It is simply removed and put somewhere else where it dissipates. All of the heat absorbed in the evaporator must be removed through the condenser for the refrigeration system to operate efficiently.

Refrigeration is used when close control of a temperature relative to a space, product or process is desired if not needed. Time is also a major factor used to determine whether refrigeration is to be used. Mechanical refrigeration speeds up the heat transfer necessary to reach a desired temperature.

Mechanical refrigeration problems stem from the operability of the equipment itself. There is quite a bit of maintenance required to ensure that the systems operate in the safest and most efficient manner possible and in accordance with regulations. Mechanical refrigeration system operations can be costly from an electrical usage or kilowatt price.

The innate problem of cost that is built-in with mechanical refrigeration can be best offset with a solid preventative maintenance and sustainability program. Making sure operators perform scheduled maintenance per the manufacturer's recommendations is essential. Sustainability efforts to keep energy use costs in line with the owner's budget goes a long way to assist in that arena.

REFRIGERATION OPERATION

Refrigeration System Operations include the safe starting, operating and shutting down of all operating equipment. Regardless of the type of system there are certain checks that an operator must follow to ensure proper operation.

In starting a refrigeration unit, the most important thing to check for first is operator safety. Systems must start and stop in the safest manner.

Operators must verify that there is power to the unit; meaning compressor and control board. But before starting a system, you must ensure that you have flow of your condensing medium so that a high head pressure does not develop as the cooling begins and the compressor sends high temperature high pressure refrigerant gas to the condenser.

All valves must be in the correct start-up position so flow is correct and in the right amount to produce the effect required.

All auxiliary systems must be ready to operate and have the power needed to do when called upon, so the system operates efficiently.

Taking Over a Shift.

Exam Prep Course (Book 2)

The following steps are necessary when entering and taking on a shift.

1. <u>Read the operations log</u>. This is the first thing that you do before actively starting your shift. An operator must first know what happened during the last shift before proceeding to even perform rounds or look at control boards or equipment. Why? Well, first it will save a trip if something is found out of the ordinary when performing equipment checks. There may be or at least should be some form of note left behind to inform the incoming operator about the prior shift's findings. Therefore, when an operator encounters an issue they will know about it and can act accordingly. The log should be kept current and include the name of all attending operators. It should list the date of operations. It should include a list of all equipment that was online during the shift and operating temperatures and pressures for every hour that rounds were performed. The log also will include any maintenance or repairs performed and where so that the incoming operators will be aware and can act accordingly.

2. <u>Performing visual rounds</u>. Once the operator has read the operations log, he/she should then walk the plant to visually inspect operations and verify the logging of the prior shift operators. This is not the time to take someone's word that "All Operations" are normal. The prior shift operators may have duplicated one round's findings and not have been back all shift. Hence, before you assume the obligation and responsibility of the attending operator you should perform your own physical and visual inspection.

3. <u>Starting a System</u>.
   a. Check valving arrangement for condenser, compressor, receiver if applicable
   b. Check flow for condenser if water cooled or a cooling tower. Start
   c. Check crankcase heater

d.   Check power to the unit and compressor drive. Start

e.   Load the compressor slowly and bring system to operating point

f.   Monitor pressures and temperatures as system comes up operating conditions

4.   <u>Shutting a System Down</u>.

a.   Shut off the compressor and stop flow.

b.   Stop water flow for condenser or cooling tower

c.   Place valves in shut-down position

d.   Monitor temperatures and pressures

e.   Check crankcase heater

5.   <u>System Preparation for Fall</u>.

a.   Each plant/system should perform a season wrap-up

b.   The system should be winterized and drained

c.   Valves will have maintenance

d.   All corrective repairs not completed should be completed at this time before next season's start-up

REFRIGERATION CYCLE

The Refrigeration Cycle is comprised of four (4) major components. When discussing the refrigeration cycle, it is the flow or direction the refrigerant takes in the system which is referred. Understanding the flow of the cycle, the direction of refrigerant flow and the state the refrigerant exists will help go a long way in furthering your comprehension for discussing on an examination.

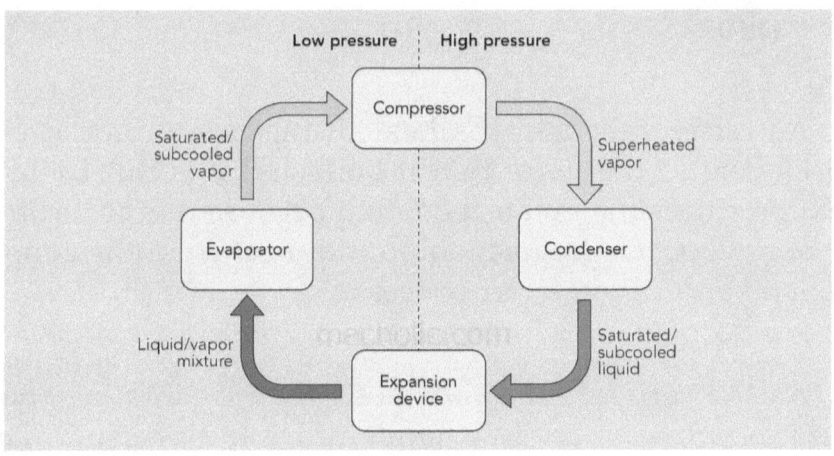

Most books begin the refrigerant cycle at the compressor. It does this because the compressor is the component that creates the pressure difference in the system which allows the refrigerant to flow. It is the compressor that pushes and pulls creating the difference in pressure that moves the refrigerant vapor or refrigerant liquid from one side of the system to the other.

All refrigerant systems are divided into two sides. One is the low side which holds the *inlet of the metering device, the evaporator,* and *the suction line leading to the compressor.* The high side is comprised of the *outlet of the compressor, the discharge line, the condenser, the liquid line, the receiver* and *the liquid line heading to the inlet of the metering device.*

I purport that the cycle begins with the evaporator. This is because refrigeration is about the removal of heat. The evaporator is the component that brings the heat into the system which is eventually removed in the condenser. If no heat is brought into the system, there can be no cooling.

As refrigerant absorbs heat in the evaporator it absorbs heat to the point that it boils and it is this boiling that produces the greatest cooling because it is absorbing or removing the greatest amount of heat from the conditioned space, product or process.

As the refrigerant vapor boils it turns into a vapor, but it is still at a very low temperature and pressure. The cool refrigerant vapor is sucked or pulled by the compressor due to the pressure differential created. At the suction valve of the compressor there is a low pressure created by the compressor.

Q. What is the refrigeration cycle?

Ans. The refrigeration cycle is the flow of refrigerant and the phases it goes through as it travels through the system and its components.

Q. What are the states of the refrigerant as it travels through the system?

Ans. On the low side of the system the refrigerant is in a low-pressure low temperature vapor. As the low temperature low pressure vapor enters the compressor its pressure and temperature are raised.

This compressing makes it a high temperature high pressure vapor leaving the compressor discharge. The compressor discharge begins the high side of the system. The high temperature high pressure refrigerant vapor leaving the compressor discharge enters the condenser and gives up its heat and changes from a vapor to a liquid in the condenser.

Leaving the condenser and traveling to or through a receive (if applicable) is a high temperature high pressure liquid. When that liquid arrives at the inlet of the metering device which is the low side of the system a portion of it flashes to a vapor because of the drop in pressure due to the reduction in pipe diameter.

The flashing cools down the evaporator surface and allows the entering refrigerant to enter as a liquid and boil as it absorbs heat from the surrounding area to be cooled. As it absorbs heat it changes from a liquid to a vapor. That vapor is at a low-pressure low temperature and is pulled through the coil to the compressor.

Q. What is meant for the low and high side of the refrigeration system?

Ans. The low side of the system is where the refrigerant vapor is at a low temperature low pressure vapor or gaseous state. The low side begins at the inlet of the metering device and ends at the suction side of the compressor.

The high side of the system is where the refrigerant is in a high temperature high pressure gaseous state or a high-pressure high temperature liquid. The high side begins at the compressor discharge and extends through to the metering device inlet.

Q. Name the major components of the refrigeration system and which side are they located?

Ans. On the low side there is the metering device, evaporator and ½ of the compressor. The high side is ½ of the compressor and the condenser, and metering device.

Q. What part does each component play in contributing to the refrigeration process?

Ans. The evaporator absorbs the heat, the compressor moves the heat, the condenser rejects the heat and the metering device feeds the evaporator with refrigerant based on the load of the system.

Q. What problems would be indicated by issues with the evaporator?

Ans. a low system pressure. Low system pressure because the evaporator is not absorbing heat into the system and thus raising the pressure and temperature of the refrigerant and affecting the capacity of the compressor.

Q. What problems would be indicated by issues with the metering device?

Ans. flooding or starving of the evaporator. Starving results from a blockage somewhere in the metering device that cause a further drop in pressure than is necessary to properly feed the evaporator. The evaporator when starved does not have the refrigerant flow through its tubes to absorb enough heat to provide the designed cooling for the system to meet load demand

When flooded the metering device is providing too much refrigerant to the evaporator. The evaporator receives so much refrigerant that all of it does not absorb heat to change its state while traveling through the coil and may send liquid refrigerant to the compressor if no accumulator is present in the suction line.

THE BEST BASIC REFRIGERATION OPERATOR

Q. What problems would be indicated by issues with the compressor?

Ans. unable to maintain desired operating pressure. The compressor's job is to move the heat or circulate the refrigerant from one side of the system to the other.

Q. What problems would be indicated by issues with the condenser?

Ans. high head or high side pressure may be a result of issues with a condenser. If the condenser is dirty or clogged inside then it's ability to reject heat is seriously impacted. And if heat cannot be rejected then its temperature and pressure will rise and possible trip out on high pressure.

METERING DEVICE

The Metering Device is where the refrigeration process begins for the most part. The metering device acts like a refrigerant club bouncer. It allows liquid refrigerant into the intake of the evaporator coil(s) based on the cooling load or amount of cooling needed to maintain the system's designed temperature.

It is the sole job of the metering device to feed the right amount of refrigerant to match the load currently on the evaporator. A faulty metering device may either flood or starve the evaporator it's connected. If there's a reduction in flow due to a restriction anywhere around the inlet, body or outlet of the metering device then the device will starve the evaporator.

This starving occurs because at the restriction or obstruction there is a drop in pressure. **Note: Where there's a drop in pressure without a drop in temperature the liquid substance will turn into a vapor.

Think of the metering device, any metering device as the evaporator's partner. The evaporator depends heavily on the metering device feeding it the right amount of refrigerant so that it can absorb all of the heat it is designed to absorb while keeping the conditioned space at the desired or set temperature regardless of load.

31

THE BEST BASIC REFRIGERATION OPERATOR

So the metering device is like the executive assistant of the evaporator. It has one job and that is to feed or let liquid refrigerant flow to the evaporator based on the load being absorbed by the evaporator. The metering device works for the evaporator. In fact, every component works for the evaporator. But this chapter is about the metering device and how it does its job, and the classifications and location.

All metering devices are located at the inlet of the evaporator. There are four classifications or types:

1. Capillary tube
2. Float
3. Thermostatic expansion

Capillary tube

Capillary tubes (cap tubes) are also used to meter refrigerant to the evaporator coil. These are considered fixed orifice since the hole in the tube(s) is a fixed size.

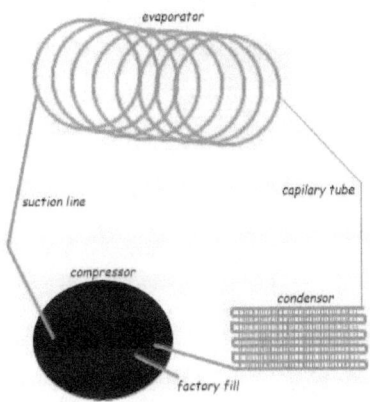

A capillary tube is nothing more than a tube that is no more than a restriction point for the liquid refrigerant to flow through to the evaporator. The restriction in diameter in the capillary tube drops the pressure of the liquid. When there is a drop in pressure of the liquid refrigerant with no corresponding drop in temperature a portion of the refrigerant flashes into vapor. This vapor helps cool the evaporator temperature so that the entering liquid does not flash and reduce the capacity of the evaporator.

The cap tube is the simplest of all the metering devices. Even though the cap tube is simple in design it still feeds refrigerant based on system load.

Thermostatic Expansion Valve

Thermostatic Expansion Valve - Also referred to as the TEV or TXV for short. The thermostatic expansion valve is used in many HVAC applications including use in chillers for chilled water systems.

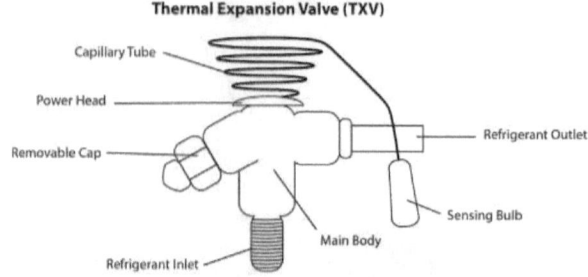

TXVs respond to the temperature of the refrigerant leaving the evaporator coil or evaporator barrel. The TXV has a sensing bulb that holds a slight refrigerant charge inside the bulb. The bulb is remote from the TXV and is attached to the TXV via a capillary tube. The bulb is attached to the suction line where the superheat leaving the evaporator coil causes the bulb to react. As the temperature increases and decreases, the refrigerant inside the bulb responds. It expands and contracts based on the temperature-pressure relationship of refrigerants. As the refrigerant expands and contracts it causes a bellows to move in and out.

This causes a piston to open and close precisely based on the leaving temperature of the refrigerant in the evaporator coil. This TXV feeds the evaporator coil the precise amount of refrigerant it needs to maintain a specific superheat. Because the TXV precisely meters the refrigerant it is used in many HVAC systems that need higher efficiency. A properly engineered and installed system that uses a TXV will only give the evaporator coil what it demands. Nothing more or nothing less in refrigerant volume except what is needed based on demand.

EXAM PREP COURSE (BOOK 2)

EVAPORATORS

The evaporator is the show of refrigeration. For years it has been told and taught that the compressor is the heart of the system. Well, that may be true since it creates the pressure difference that allows the refrigerant to flow from one side to another but without heat being absorbed into the system there is in fact no refrigeration occurring ever. The compressor moves the heat but only after the evaporator brings it into the system.

The evaporator is a coil where the heat is absorbed into the system thus cooling the area around it. The liquid refrigerant boils and turns to a gas and pulled into the suction line. Because of the way they operate, evaporator and condenser coils both need to be kept clean to perform as intended and reach optimal energy efficiency.

Different types of evaporators are used in different types of refrigeration applications and accordingly they have different designs. The evaporators can be classified in various ways depending on the construction of the evaporator, the method of feeding the refrigerant, the direction of circulation of the air around the evaporator, etc.

Classification of the Evaporators Based on the Construction

Exam Prep Course (Book 2)

The evaporators used for refrigeration and air conditioning applications have different types of construction depending on the application. Based on their construction the various types of evaporators are:

1. 1.   Bare Tube Evaporators

The bare tube evaporators are made up of copper tubing or steel pipes. The copper tubing is used for small evaporators where the refrigerant other than ammonia is used, while the steel pipes are used with the large evaporators where ammonia is used as the refrigerant. The bare tube evaporator comprises of several turns of the tubing, though most commonly flat zigzag and oval trombone are the most common shapes.

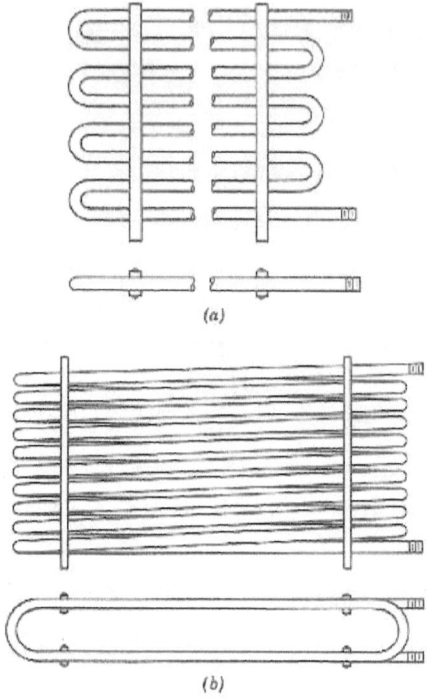

(a)

(b)

The bare tube evaporators are usually used for liquid chilling. In the blast cooling and freezing operations, the atmospheric air flows over the bare tube evaporator and the chilled air leaving it used for cooling purposes. The bare tube evaporators are used in very few applications, however the bare tube evaporators fitted with the fins, called as finned evaporators are used very commonly.

2. Plate Type of Evaporators

In the plate type of evaporators, the coil usually made up of copper or aluminum is embedded in the plate so as to form a flat looking surface. Externally the plate type of evaporator looks like a single plate, but inside it there are several turns of the metal tubing through which the refrigerant flows.

The plate type heat exchangers can be easily formed into various shapes as per the requirement. Thus, in the household refrigerators and deep freezers, where they are used most commonly, they can be converted into a box shape to form the closed enclosure, where various food can be kept in the frozen state. The plates can also be welded together forming the bank of the plate type of evaporators that can be used the larger evaporators of higher capacities.

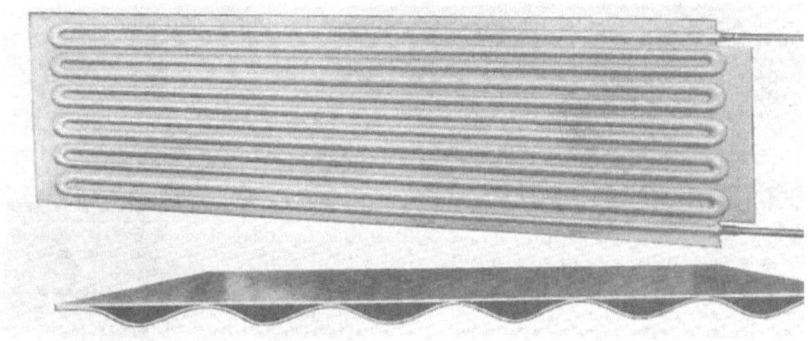

The advantage of the plate type of evaporators is that they are more rigid as the external plate provides lots of safety. The external plate also helps increasing the heat transfer from the metal tubing to the substance to be chilled. Further, the plate type of evaporators is easy to clean and can be manufactured cheaply.

3. Finned Evaporators

The finned evaporators are the bare tube type of evaporator covered with the fins. When the fluid (air or water) to be chilled flows over the bare tube evaporator lots of cooling effect from the refrigerant goes wasted since there is less surface for the transfer of heat from the fluid to the refrigerant. The fluid tends to move between the open spaces of the tubing and does not meet the surface of the coil, thus the bare tube evaporators are less effective. The fins on the external surface of the bare tube evaporators increases the contact surface of the metallic tubing with the fluid and increase the heat transfer rate, thus the finned evaporators are more effective than the bare tube evaporators.

The fins are the external protrusions from the surface of the coil, and they extend into the open space. They help removing the heat from the fluid that otherwise would not have meet the coil.

For the fins to be effective it is very important that there is a very good contact between the coil and the fins. In some cases, the fins are soldered directly to the surface of the coil and in other cases the fins are just slipped over the surface of the fins and then they are expanded thus ensuring a close thermal contact between the two. Tough the fins help increase heat transfer, rate, adding them beyond certain numbers won't produce any additional benefits, hence only certain number of fins should be applied on the external surface of the tube.

The finned evaporators are most commonly used in the air conditioners of almost all types like window, split, packaged and central air conditioning systems. In these systems the finned evaporator is popularly known as the cooling coil. The hot room air flows over the finned evaporator or the cooling coil, gets chilled and enters the room to produce the cooling effect. In window a/c the finned evaporators are located behind the beautifully looking grill. In the wall mounted split unit it located behind the front grill of the indoor unit.

Q. What are some issues that affect evaporator performance?

Ans. A dirty evaporator coil can experience several problems, including:
- Impaired heat absorption and cooling capacity
- Higher energy use
- Higher pressures and temperatures
- Frost and ice buildup

Q. How does a dirty evaporator affect system performance?

Ans. Even a fine layer of dust on the evaporator coil reduces its efficiency. The dust acts as an insulator, keeping the heat in and the air away from the cold coils. That means the coil can't absorb as much heat as it can when clean. Your system will then have to run longer to provide the indoor temperature you want, which means it will use more energy.

Because it isn't absorbing enough heat, the refrigerant running through a dirty evaporator coil doesn't warm up as much as it should. This very cold refrigerant causes water vapor in your air to freeze rather than condense into a liquid. Eventually, the whole evaporator coil can frost over.

Q. Is a layer of frost on an evaporator coil normal?

Ans. A layer of frost on your evaporator is never normal. Letting your system run with a frozen evaporator raises the temperature in the compressor and can eventually cause this component to fail. Dust on the evaporator coil, debris on the outdoor condenser unit, a dirty filter, and a refrigerant leak can all cause the evaporator to freeze. If you can't pinpoint the problem, contact a heating and cooling technician.

Q. Can evaporator coils develop leaks?

Ans. Evaporator coils can also develop tiny pinhole leaks due to corrosion caused by the mixing of moisture from condensation with chemicals commonly found in household air. Oily residue around the evaporator or in the drain pan is a sign your coil is leaky and requires replacement.

Q. What are VOCs?

Ans. The airborne chemicals that encourage these

leaks are known as volatile organic compounds (VOCs) and come from new carpeting, upholstery, pressed wood furniture, air fresheners, cleaning chemicals and many other sources. Ensuring good ventilations reduces the VOCs in your indoor air, protecting both the evaporator coil and your health.

COMPRESSORS

The component which creates the pressure difference in the system thus creating the flow of the refrigerant pulling it from the evaporator and pushing it through the condenser and back to the metering device. Without it no flow would occur and thus no cooling. The compressor takes the low temperature low pressure refrigerant gas and compresses it. The temperature and pressure raise so it can give off heat in the condenser and turns back into a liquid to be metered back into the evaporator by the metering device.

Q. What is a compressor?

Ans. A refrigeration compressor is a refrigerant gas

pump in which the evaporator supplies gaseous refrigerant at a low pressure and increases it to a greater pressure. Upon being compressed, the temperature and pressure of the vapor are increased. The gaseous refrigerant is delivered to the condenser at a pressure at which condensation occurs at an appropriate temperature.

Q. What are some factors that affect compressor performance?

Ans. Factors which affect the performance of compressors are:
- speed of rotation
- pressure at suction
- pressure at discharge and
- type of refrigerant being used

Similar compressors can operate at different capacities by varying their refrigerants and compressor horsepower input. When purchasing any type of compressor, the buyer should check certain characteristics that include the machine configuration, the operation type, the price, and the operating cost. In any case, he should check the performance of the compressor and consult with the manufacturer about the most suitable and safest compressor for his budget and requirements.

Q. Name the types of compressors used in refrigeration.

Ans. There are basically 5 types of air conditioner compressor that are commonly used in the HVAC industry:

- Reciprocating.
- Scroll.
- Screw.
- Rotary.
- Centrifugal.

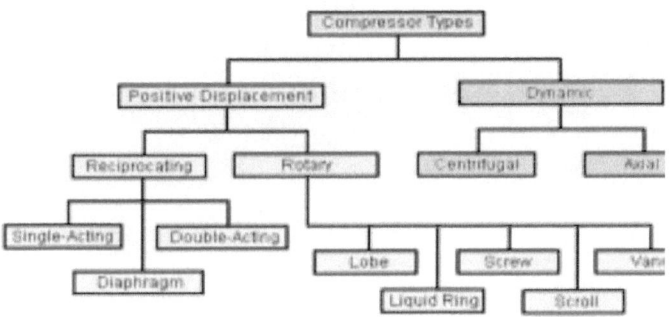

The image above shows the available types of compressors. The most common ones used in refrigeration are described below:

Q. Describe a rotary compressor.

Ans. Compressors of the rotary type are generally low capacity equipment, used normally in home refrigerators and freezers, and not used for air conditioning. These compressors can consist of one vane, which is placed in the body, and sealed against the rotor, or multi-vane rotary, with vanes located in the rotor.

Q. Describe a Centrifugal Compressor.

Ans. These compressors revolve at high speed, and refrigerant is compressed by the application of centrifugal force. These compressors are normally used with refrigerants possessing higher specific volumes, which require lower compression ratios. Multi-stage units can be used to attain greater discharge pressures, and the number of stages is determined by the discharge temperature of the gas as it exits from the rotor. These compressors are utilized for water chilling in air conditioning and for low temperature freezing purposes.

Q. What is a Reciprocating Compressor?

Ans. These compressors have pistons and move in cylinders.

Q. What are t*he types of reci*procating compressors?

Ans. Open Compressors: One extremity of the crankshaft is drawn out of the crankcase, due to which multiple drives can be used with the compressor. A mechanical seal is used to check external seepage of refrigerant and oil, and escape of air towards the inside. These compressors are driven by electric motors or internal combustion engines. With belt drive, changes in speed are achieved by altering the dimensions of the pulleys, while with direct drive units the compressor *is planned to operat*e at the speed of motor.

Hermetic Compressors: These compressors are serviceable hermetic, in which motor and compressor are enclosed in the same housing, while the welded hermetic type has the compressor and motor sealed in a welded steel shell.

CONDENSERS

A coil similar to the evaporator but larger in size so that it can be sure to reject the heat absorbed into the system by the evaporator plus the heat added to the refrigerant by the compressor and suction line and discharge lines. The condenser is where the refrigerant turns from a gas back to a liquid for reuse.

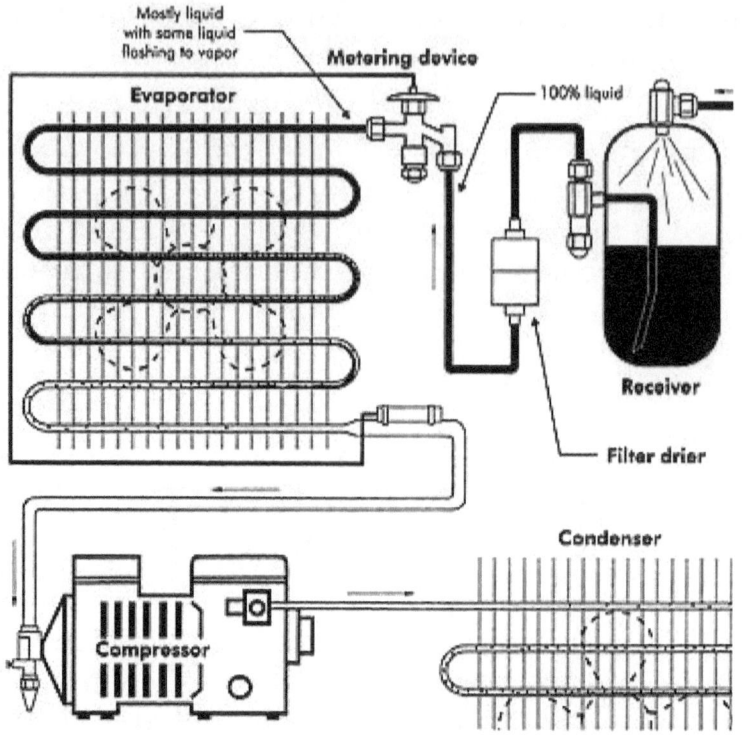

Condensers are the same device as an evaporator. The major difference is that it is slightly larger due to its requirement to remove all of the heat the evaporator brings in plus the heat of compression. Yet, overall it is still just a heat exchanger device.

Condensers are sized at 20-25% larger than evaporators. Condensers are designed to discharge liquid refrigerant to the liquid line or receiver (if applicable). If the condenser is unable to reject heat as designed the high side pressure will rise backing up and slowing down refrigerant flow being pumped from the compressor. This eventually rises pressures on the low side and prevents the evaporator from turning liquid into a vapor since a greater portion of the evaporator surface being already covered with vapor.

Function of Condenser

In a cooling cycle of a refrigeration system, heat is absorbed by the vapor refrigerant in the evaporator followed by the compression of the refrigerant by the compressor. The high pressure and high temperature state of the vapor refrigerant is then converted to liquid at the condenser. It is designed to condense effectively the compressed refrigerant vapor.

There are basically three types of condensing unit depending on how the heat is removed by the condensing medium which is usually water, air or a combination of both.

- Air-Cooled types are usually used in the residential and small office applications. They are used in small capacity systems below 20 tons. The advantages of using this design include not having to do water piping, not necessary to have water disposal system, saving in water costs and not much scaling problems caused by the mineral content of the water. It is also easier to install and has a lower initial cost. There aren't much maintenance problems. The disadvantages are that it requires higher power per ton of refrigeration, has shorter compressor life and on days when most cooling is required, the least is available.

Though aesthetically not pleasant, many installers use aluminum duct to redirect the hot air away from the condenser.

- The circulation of air-cooled type can be by natural convection or by forced convection (usually using blower or fan). Due to its limited capacity, natural convection is used in smaller applications such as freezers and refrigerators. In forced convection, air is circulated by using a fan or blower that pulls the atmospheric air through the finned coils. Internally, the refrigerant circulates through the coil and air flows across the outside of the tubes.

- Water-Cooled There are 3 types commonly used. They are *shell and tube*, *shell and coil*, and *double tube*. The most commonly used is the shell and tube type and are usually available from two tons up to a couple of hundred tons. This design has lower power requirements per ton of refrigeration and the compressors can last longer compared to the air-cooled type. A water-cooling tower is frequently used for higher capacity applications.

- Evaporative type which is a combination of water and air-cooled.

Air-Cooled and Water-Cooled Comparison Summary

- Air-cooled type operates at higher head pressure or condensing pressure, hence reducing the capacity of the compressor and increases the power intake. In general, a 2 hp water-cooled system will require the same refrigeration as a 3 hp air-cooled system.

- The maintenance costs of water-cooled type is about three to four times the air-cooled type. Air-cooled type maintenance is usually limited to regular lubrication of fan and motor bearings. Water-cooled type requires cleaning from algae and bacteria. Scales on the tubes are removed by using acid compound. Proper water treatment is also critical to the operation of the condenser.

Exam Prep Course (Book 2)

RECEIVERS

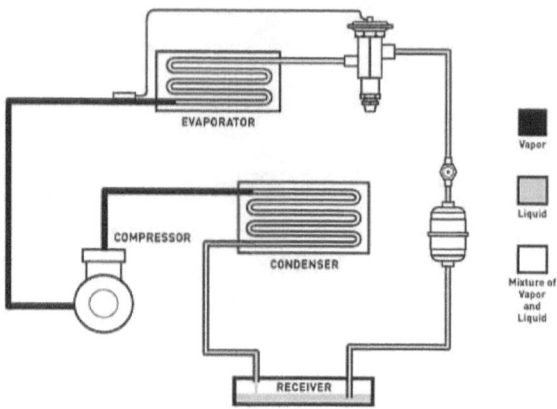

A common accessory used on many refrigeration systems is the liquid receiver. It is basically a storage vessel designed to hold excess refrigerant not in circulation. Refrigeration systems exposed to varying heat loads, or systems utilizing a condenser flooding valve to maintain a minimum head pressure during low ambient temperatures, will need a receiver to store excess refrigerant.

Liquid receivers are installed in the liquid line as close as possible to the outlet of the condenser. The piping between the condenser and the receiver should be arranged to allow free drainage. The piping should also not cause excessive friction pressure loss or gas binding and must have adequately sized valves and connection fittings.

The location of the receiver should not cause excessive heat to be added to the refrigerant, such as from direct solar radiation when located outdoors or near building heating equipment when installed indoors. Excess heat added to a receiver will reduce the operating efficiency of the system. However, if the receiver is installed outdoors and the system is required to operate during low ambient temperatures, it may be necessary to install trace heaters to maintain adequate pressure in the receiver in order to avoid system problems at startup.

Some receivers are built into the condenser. The receiver has these advantages:

- It eliminates the need for an exact refrigerant charge
- It is a handy place to store refrigerant during servicing
- It helps prevent flash gas by ensuring that liquid refrigerant remains subcooled
- It can hold the refrigerant during automatic pumpdowns, such as for defrosting or when some evaporators shut off

To prevent overpressure in a receiver, a pressure or temperature sensitive safety device must be connected to a vent line to the outside.

SYSTEM ACCESSORIES

Purger

The purger system removes undesirable gases (air) from the system to enhance the operating efficiency of the compressors and condensers. Regardless of the type of refrigerant used, removing air quickly and efficiently is essential. Air finds its way into any refrigeration system, no matter how carefully it is maintained. Air enters the system in several ways:

- Leaking through seals and valve packing when suction pressure is below atmospheric conditions
- When the system is open for repairs, coil cleaning, and equipment additions
- When refrigerant trucks charge the system
- When oil is added
- Through the breakdown of refrigerant or lubricating oil
- From impurities in the refrigerant.

Why remove air?

Insulating properties of air are well known. Air molecules generated in the gas by the compressor accumulate on the inner heat transfer surface of the condenser. This accumulated air both insulates the transfer surface and effectively reduces the size of the condenser. (A good analogy is cholesterol and fatty deposits clogging arteries.) To offset this size reduction, the system must work harder by increasing the pressure and temperature of the refrigerant.

Air in the system typically causes excessive wear and tear on bearings and drive motors and contributes to a shorter service life for seals and belts. Plus, the added head pressure increases the likelihood of premature gasket failures. The most obvious reason to remove air is evident on the utility bill. For each 4 lb of excess head pressure caused by the air, the power cost to operate the refrigeration system compressor increases 2% and the compressor's capacity drops 1%. This reason alone makes it essential to choose the proper size and type of purger for the job.

Air in the system

The easiest way to determine the amount of air in a refrigeration system is to check the condenser pressure and the temperature of the refrigerant leaving the condenser. Then, compare the findings with the data found in a temperature-pressure chart (See Table I for an abbreviated chart. The complete table is contained in the ASHRAE 1997 *Fundamentals Handbook*. Information for obtaining the handbook is found in the *More info* box at the end of this article.)

For example, if the ammonia temperature is 86 F, the theoretical condenser pressure should be 154.5 psig. If the gauge reads 174 psig, the 20-psi excess pressure is increasing power costs 10% and reducing compressor capacity 5%.

Performing a purge

THE BEST BASIC REFRIGERATION OPERATOR

Air is removed from a system two ways: manually or automatically. When a system is purged manually, first a valve is opened by hand to let the air escape. Seeing a cloud of refrigerant gas discharging from the system does not mean the system has been purged. Until the mechanical purger was introduced in 1940, manual purging was the only option available. However, manual purging wastes refrigerant, takes a lot of time, and does not totally eliminate air. It permits an escape of refrigerant gas that may be dangerous and disagreeable to people and the environment. Because of the drawbacks, manual purging is often neglected until the presence of air in the system causes problems. Therefore, automatic purging is preferred.

Choosing an automatic purger

Determining which automatic purger to use depends primarily on whether power is available at the purger location and safety considerations permit the use of electrical components. Let's consider two types of automatic purgers: nonelectrical automatic mechanical types and automatic electronic refrigerated types (single point and multipoint).

Nonelectrical automatic mechanical units are used primarily when there is no electricity at the point of use or in hazardous applications where electronic components are not allowed. These units remove non-condensable gases from refrigeration systems by determining the density difference between the liquid refrigerant and gases. An operator opens and closes valves to start and stop the purging operation and ensure its efficiency.

Automatic electronic refrigerated purgers offer additional benefits when conditions permit their use. There are two types of electronic purgers: single point and multi-point. A single point unit performs a mechanical purge operation with a temperature/gas level monitor controlling the discharge to atmosphere. The purging sequence is done manually or tied to a PLC.

The multi-point purger handles several points from the same unit. However, each point is individually purged. The multi-point purger offers total automation and includes start-up, shutdown, and alarm features. This type of purger *must* be designed for the total tonnage of the system. Small purgers may cost less initially but may adversely impact system efficiency and ultimately the payback period.

The most recent generation of multi-point purger includes a microprocessor based, fully programmable controller. The controller learns as it cycles through the system. As the purger accumulates air and purges, the controller records and prioritizes each purge point in its memory, thus removing air more efficiently.

Locating purge points

Before air is removed from a system, the locations where it is likely to accumulate or collect inside the system must be identified. Multiple condensers and receivers make it difficult to determine the exact location of the air. Condenser piping design and component arrangement and operation affect the location of air. Seasonal weather changes also affect air location. Therefore, it is important to purge each purge point regularly and frequently one point at a time to ensure that all the air is removed from every possible location.

As a rule, refrigerant gas enters a condenser at a high velocity, but by the time it reaches the far (and cool) end of the unit, its velocity is practically zero. This point is where air accumulates and where purge points need to be made. Similarly, the purge point connection on the receiver should be made at the point farthest from the liquid inlet. Always locate the purge connection at the top of the pipe and above the discharge point of the liquid refrigerant.

Oil Separator

Oil separator minimizes the amount of oil circulating in the cooling system. Oil to coat the inside of each cycle of the cooling components, through which it passes. This is one of the reasons for separation as much oil as possible with the refrigerant before sending it to the system. Another reason is to slow down the accumulation of oil in those places where it is difficult to return the oil. This consideration is particularly important where the refrigerant is not mixed. The refrigerant does not dissolve in oil and oil is not soluble in refrigerant.

Oil separators are recommended for all systems using non-miscible refrigerants. Ammonia is not mixed, and R-11, R-12, R-113, R-114, R-500 and mixed. R-22 and R-500 are limited compatibility but mixed with oil for normal comfort air conditioning temperatures. Oil separators are also recommended for low-temperature applications and flooded evaporators.

Finally, oil separators are used at any split-system or system with suction or discharge struts and/or capacity reduction steps. These functions are often the cause of problems with oil return. In practical terms, this means that oil separators are usually not found on residential and commercial comfort air conditioning systems.

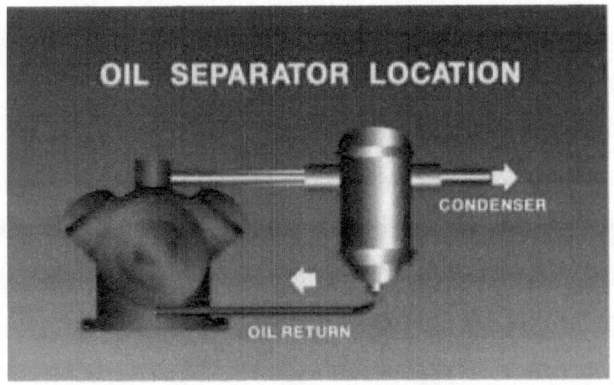

The only goal of oil in the refrigeration cycle is lubrication of the compressor's moving parts. Thus, oil separator installed in such a way that oil will be chained to the compressor. It is usually installed in the discharge line from the compressor as close as possible. If a silencer on the compressor is used then it is set after the silencer, if the silencer. Oil return line is connected directly to the compressor crankcase in some designs. In others, it is forwarded to the compressor suction pipe.

Although oil separators, when used correctly, are very effective in the removal of oils with the refrigerant, they are not able to delete the circulation of oil with gas discharge. Some oils still find its way into the system. Therefore, when oil separators are used, some means must be provided periodically return the accumulated oil in the compressor. Oil flows are provided for this purpose at the bottom of the receivers, condensers, evaporators, batteries, and other vessels used for ammonia systems. Automated methods are used.

Compressor oil meets the refrigerant in the compressor. As a result, a certain amount of oil will be carried out along with the compressed refrigerant leaves the compressor through the discharge line.

Crankcase heaters

Many air conditioning and refrigeration systems have their condensing units located outdoors for two main reasons. First, this takes advantage of the cooler outdoor ambient temperatures to reject the heat absorbed in the evaporator section, and second, to reduce noise pollution.

The condensing unit usually consists of a compressor, condenser coil, outdoor condenser fan, contactor, a starting relay, capacitors, and a solid-state board with circuitry. Receivers are often incorporated into refrigeration system condensing units. Within the condensing unit, the compressor will often have a heater connected in some way to its bottom section or crankcase. This heater is often referred to as a crankcase heater.

The crankcase heater is an electric resistance heater that is usually strapped on, clamped to the crankcase bottom, or inserted into a well within the compressor's crankcase. The crankcase heater is often seen on compressors that operate in ambient temperatures lower than the system's operating evaporator temperature.

The compressor's crankcase lubricant, or oil, has

many important functions. Even though the refrigerant is the working fluid required for cooling, oil is needed to lubricate the compressor's moving mechanical parts. Under normal conditions, there will always be a small amount of oil that escapes a compressor's crankcase and circulates with the refrigerant throughout the system. The proper refrigerant velocity traveling through the system's tubing will return this escaped oil to the crankcase over time, and it is for this reason that oil and refrigerant must be soluble in one another. At the same time, however, the solubility of the oil and refrigerant can cause another system problem. That problem is refrigerant migration.

Migration is an off-cycle phenomenon. It is a process where either liquid and/or vapor refrigerant migrates or returns to the compressor's crankcase and suction line during the compressor's off cycle. During a compressor's off cycle, and especially during a long shutdown period, refrigerant will want to travel, or migrate, to a place where the pressure is the lowest. In nature, fluids travel from a place of higher pressure to a place of lower pressure. The crankcase usually has a lower pressure than the evaporator because of the oil it contains. A cold ambient temperature will amplify the lower vapor pressure phenomenon and help condense the refrigerant vapor to a liquid inside the crankcase.

Refrigerant oil itself has a low vapor pressure and will flow to it whether the refrigerant is in the vapor or liquid state. In fact, refrigerant oil has such a low vapor pressure, it will not vaporize even when a 100-micron vacuum is pulled on the refrigeration system. Some refrigeration oils have a vapor pressure as low as 5-10 microns. If the oil did not have such a low vapor pressure, it would vaporize every time a low pressure existed in the crankcase or when a vacuum was pulled on it.

Because refrigerant migration can occur with refrigerant vapor, the migration can occur either uphill or downhill. When the refrigerant vapor reaches the crankcase, it will be absorbed and condense in the oil because of the refrigerant/oil miscibility.

On long off cycles, the liquid refrigerant will be on the bottom of the oil, forming a striated layer in the crankcase. This happens because liquid refrigerant is heavier than oil. On short compressor off cycles, the migrated refrigerant does not have a chance to settle under the oil but will still mix with the oil in the crankcase.

Operators often turn off the electrical disconnect to their air conditioner's outdoor condensing unit during the heating season and/or cooler months when air conditioning is not needed. This will cause the compressor to be without crankcase heat because of a de-energized crankcase heater. Migration of refrigerant to the crankcase is sure to take place during this long off cycle.

Once the cooling season begins, if the operator doesn't turn the breaker back on at least 24-48 hours before starting the air conditioning unit, serious crankcase foaming and pressurization will occur from the long off-cycle refrigerant migration.

This can rob the crankcase of a proper oil level as well

as score bearings and cause other mechanical failures within the compressor.

The crankcase heater is designed to help combat refrigerant migration. The function of the crankcase heater is to hold the oil in the compressor's crankcase at a temperature higher than the coldest part of the system. This will cause the crankcase to have a slightly higher pressure than the rest of the system. Refrigerant entering the crankcase will then be vaporized and driven back into the suction line.

Migration of refrigerant to the compressor's crankcase during an off cycle is a serious problem. Severe compressor damage can result if the problem is not remedied. If refrigerant migration does occur, when the compressor starts for its next on cycle, there will be an immediate drop in crankcase pressure from the startup. This will cause violent foaming in the crankcase. The striated refrigerant and oil mixture will explode, causing rich refrigerant/oil foam to form in the crankcase. This pressurized refrigerant and oil foam will seep through the cavities of the compressor, causing broken discharge valves and reeds. The oil level in the crankcase will then drop, and mechanical parts will be scored from inadequate lubrication. With the high crankcase pressure, the mixture of refrigerant and oil foam can now be forced around piston rings and be pumped by the compressor into the refrigeration system. Not only does this situation cause loss of oil from the crankcase, it also can cause a mild form of oil and/or refrigerant slugging in the compressor's cylinders.

Slugging is when liquid refrigerant or liquid refrigerant and oil enter the compressor's cylinder during the compressor's on cycle. High compressor current draw, which will lead to motor overheating, usually follows. Also, broken or warped valves can occur as a result of overheating and/or slugging.

The only sure remedy for refrigerant migration to a compressor is an automatic pump-down system. One must clear all of the refrigerant from the evaporator and suction line before every off cycle. Automatic pump-down is accomplished with a thermostat controlling a liquid line solenoid in combination with a low-pressure controller terminating the on cycle once the evaporator and suction lines are void of any refrigerant. This will ensure there is no refrigerant in the evaporator or suction line to migrate toward the compressor.

Some control schemes pump down the evaporator and suction line before each off cycle and, at the same time, energize a crankcase heater during the off cycle. Others employ both a crankcase heater and a properly sized suction line accumulator to protect the compressor from liquid returning to the compressor. However, in severe cases, a suction line accumulator can also be flooded with refrigerant.

Crankcase heaters can be energized continuously or during the on cycle. However, in order to avoid carbonizing of the oil from excessive heat, the wattage input of the crankcase heater must be limited. In ambient temperatures approaching 0°F, or when exposed to cold winds, the crankcase heater may be overpowered, and refrigerant migration to the compressor's crankcase may still occur.

The crankcase heater is also a safety precaution in case the liquid line solenoid on an automatic pump-down system leaks refrigerant during the off-cycle. The crankcase heater will prevent any refrigerant from getting to the crankcase and causing oil flash at start-up. However, it will not prevent slugging or flooding of liquid refrigerant from the suction line or evaporator at startups. This is because the next compressor on cycle could draw this refrigerant from the suction line, and liquid slugging of the compressor can occur. Again, crankcase heaters do help in combating refrigerant migration to the compressor's crankcase but do not prevent compressor slugging at startups or liquid floodback to compressors once the compressor is running.

It is often thought that a crankcase heater will prevent migration. Crankcase heaters will keep the compressor's crankcase warm and prevent refrigerant migration to the compressor's oil in the crankcase. However, condensed migrated refrigerant can sit in the suction line near the compressor, waiting for the next on-cycle. If excessive liquid refrigerant has been driven to the suction line, severe liquid slugging may occur during startups. Frequently, compressor damage, such as broken valves and damaged pistons, will occur.

Crankcase heaters can be effective in combating refrigerant migration to the compressor's oil in the crankcase, but they will not remedy slugging at startups from liquid floodback unless used in conjunction with a properly sized suction line accumulator. And, the only sure way to prevent refrigerant migration is with an automatic pump-down system.

Q. What is the purpose of a crankcase heater?

Ans. The crankcase heater is an electric resistance heater that is usually strapped on, clamped to the crankcase bottom, or inserted into a well within the compressor's crankcase. The crankcase heater is often seen on compressors that operate in ambient temperatures lower than the system's operating evaporator temperature.

Q. Where is the crankcase heater located?

Ans. On the low side of the system.

Q. How does a crankcase heater work?

Ans. The crankcase heater is designed to help combat refrigerant migration. The function of the crankcase heater is to hold the oil in the compressor's crankcase at a temperature higher than the coldest part of the system. This will cause the crankcase to have a slightly higher pressure than the rest of the system. Refrigerant entering the crankcase will then be vaporized and driven back into the suction line.

Migration of refrigerant to the compressor's crankcase during an off cycle is a serious problem. Severe compressor damage can result if the problem is not remedied. If refrigerant migration does occur, when the compressor starts for its next on cycle, there will be an immediate drop in crankcase pressure from the startup. This will cause violent foaming in the crankcase. The striated refrigerant and oil mixture will explode, causing rich refrigerant/oil foam to form in the crankcase. This pressurized refrigerant and oil foam will seep through the cavities of the compressor, causing broken discharge valves and reeds. The oil level in the crankcase will then drop, and mechanical parts will be scored from inadequate lubrication. With the high crankcase pressure, the mixture of refrigerant and oil foam can now be forced around piston rings and be pumped by the compressor into the refrigeration system. Not only does this situation cause loss of oil from the crankcase, it also can cause a mild form of oil and/or refrigerant slugging in the compressor's cylinders.

Liquid Line Sight Glass

Sight glass-window that allows the mechanic to see inside the system. This allows the installer or Servicer to observe the state of the refrigerant in the sight glass in place. Chemical compounds on the drive sits in the refrigerant flow and change color in the presence of water. The amount of change indicates the amount of water.

Q. Will a sight glass inform you if the system is over-charged?

Ans. Each field rechargeable system must have a means to verify the enough the refrigerant. Correctly located sight glass meets this need. It will not, however, tell you if the system is overcharged. Sight glass should have a viewing window on either side (dual port), so a flashlight may be shone through the refrigerant. In addition, sight glass should have a seal cap on every port to prevent leakage.

In this place, sight glass has an additional function, which shows a pure liquid enters the metering device, as it should. When the pressure drops from the entering a metering device is substantial, additional sight glass must be installed at the outlet.

Q. When a refrigeration system has a correct charge what should the operator see in the sight glass?

Ans. When the charge is correct, but the bubbles appear before the metering device, it means the flashing occurs in liquid line due to pressure drop. This situation can be eliminated by reducing the pressure drop for further sub-cool fluid.

Liquid line drier

Filter-driers are a key component in any refrigeration or air conditioning system. A filter-drier in a refrigeration or refrigeration system has two essential functions:

1. To adsorb system contaminants, such as water, which can create acids, and

2. To provide physical filtration. Evaluation of each factor is necessary to ensure proper and economical drier design.

The ability to remove water from a refrigeration system is the most important function of a filter drier. Water can come from many sources, such as trapped air from improper evacuation, system leaks, and motor windings, to name a few.

Another source is due to improper handling of polyolester (POE) lubricants, which are hygroscopic; that is, they readily absorb moisture. POEs can pick up more moisture from their surroundings and hold it much tighter than the previously used mineral oils. This water can cause freeze-ups and corrosion of metallic components.

Water in the system can also cause a reaction with POEs called hydrolysis, forming organic acids.

Q. How are the formation of acids prevent or minimized using desiccants?

Ans. To prevent the formation of these acids, the water within the system must be minimized. This is accomplished by the use of desiccants within the filter drier. The three most commonly used desiccants are molecular sieve, activated alumina, and silica gel.

Q. What factors are involved in selecting a desiccant material?

Ans. There are many factors involved when selecting which desiccant material is best for an application. Water capacity, refrigerant and lubricant compatibility, acid capacity, and physical strength are important characteristics of desiccants and should be considered. The first of these, water capacity, is the amount of water the desiccant can hold while maintaining low moisture levels within the refrigeration system.

A molecular sieve retains the highest amount of water, while keeping the concentration of water in the refrigerant low. This is due to the strong bond between the molecular sieve and the water.

Q. How is freeze-ups and corrosion minimized with filter-driers?

Ans. By keeping the water in the system at low levels, freeze-ups, corrosion, and acid formation is minimized. Activated alumina retains a fair amount of water, but the retention isn't as great as the molecular sieve. This is indicative of co-adsorption of other material. Based on this information, Parker recommends the use of 100% molecular sieve in liquid line filter-driers for maximum water removal.

Q. What should be considered in determining refrigerant and lubricant compatibility?

Ans. Refrigerant and lubricant compatibility is also

essential when selecting a desiccant. Inorganic acids (HCl and HF) form from the decomposition of the refrigerant reacting with an incompatible desiccant and water at elevated temperatures. Inorganic acids formed will attack the crystalline structure of the molecular sieve and break it down as well as attack metal surfaces in the system. Organic acids can form from the breakdown of the lubricant in the presence of an incompatible desiccant and water (elevated temperatures will increase this reaction).

These organic acids are a sludge-like material that can deposit and plug the system's expansion device. Parker has tested each of the desiccants used based on their application, to ensure that the formation of these acids is minimized. The varying pore sizes in the activated alumina allow it to be more effective than molecular sieve in removing the larger, organic acid molecules.

Alumina is more effective in removing the various acids when it is used in the suction line of the system. When used in the liquid line of a system, there is a potential for the hydrolysis reaction between the lubricant and water to occur, forming organic acids. This reaction did not occur when the alumina was tested in the suction line. Therefore, for acid cleanup in a system, some manufacturers recommend the use of a suction line filter-drier containing an activated alumina core.

Q. What factors of a desiccant should be considered?

Ans. Physical strength of the desiccant is another factor to be considered. Desiccants should be strong enough mechanically to resist breaking up when subjected to system vibrations and surges (attrition). Attrition occurs when the desiccant beads rub against one another when it is shaken or vibrated, yielding fine particles. Therefore, the method of retaining the desiccant in the filter-drier (based on drier size and location) plays a major role on the integrity of the desiccant.

Q. What is the main function of a filter-drier?

Ans. Filtration is the other main function of a filter-drier and is accomplished by different methods. Some filter driers use only one method; others may use a combination of methods.

Q. What are the two-primary means of mechanical filtration?

Ans. There are two primary means of mechanical filtration: *surface and depth*.

The simplest form of surface filtration is the screen. The screen is usually a woven wire mesh that catches particles that are larger than the holes in the screen. Until the screen has captured enough particles to provide a layer across the entire surface, particles that are smaller than the holes will pass through the screen. In addition, a particle longer than a hole can pass through if its cross-section is smaller than the hole.

As layers of contaminant cover the screen, it will become a depth filter as the layer of contaminant will act as a filter to remove smaller particles that would ordinarily pass through the screen. This layering of contaminant will continue until the pressure drop across the screen reaches the point at which the refrigerant flashes into vapor.

Depth filtration takes different forms.

Q. What are the most common depth filters?

Ans. The most common depth filters are:

- Bonded desiccant cores
- Rigid fiberglass filters bonded with phenolic resin; and
- Fiberglass pad filters.

Depth filters force the fluid and contaminant to take an indirect route through the filter. Contaminants are trapped in a maze of openings that are spread throughout the filter. Depending on the type of filter, the openings will vary significantly.

Bonded desiccant cores have smaller rigid openings than do fiberglass pads. As the flow passes through the media, particles are trapped in the channels, depending upon their size. As the channels fill with particles, the pressure drop will increase until vaporizing occurs as described above.

Fiberglass pad filters are not compressed as tightly as bonded, rigid fiberglass filters. The liquid refrigerant with the entrained contaminant flows through the pads. The contaminant will impact the glass fibers and lose some velocity. As the contaminant passes through the media, the velocity will eventually drop to zero, at which point the contaminant will deposit in an opening in the fiberglass. The larger particles will tend to drop out first as their higher mass will tend to cause them to impact on a fiber even though the flow stream will bend around a fiber.

As the fiber glass fills with more and more particles, the filtration becomes finer as the filter becomes closer in function to the rigid filter. The core drier picks up particles and the pressure drop increases quickly as the core plugs with contaminant. For the same pressure drop and flow rate, the fiber pad drier can hold up to five times the amount of contaminant as the core drier with equivalent or greater filtration capacity.

The core can be used effectively in the suction line drier. In this case, the higher velocity in the suction line will cause the loose fiberglass structure to disintegrate. The rigid cores can be tailored to remove the solid particles that result from compressor breakdown, sludge, and resins.

The desiccant bonded in the core will remove water and neutralize acids caused by breakdown of the lubricant. The bonding of the desiccant will preclude the attrition that can occur with loose desiccant beads.

REFRIGERANT CONTROLS

There are only a few controls on a normal Refrigeration system. To be a good operator you must know what they are, what they do and when they do it. You need to know this in order to know how they function in conjunction with the major components they serve.

The most common refrigeration control is the Thermostat. Sounds simple.

Q. What are some of the functions of the thermostat in refrigeration systems?

Ans.

1. Thermostats cut the compressor off.
2. Thermostats cut a solenoid valve off and the Pressure Control cuts the compressor off after the system pumps down.
3. Controls the time clock that works with the defrost cycle
4. Controls the defrost system so the coils are free of

frost before starting up

Q. What are some of the functions of a dual pressure control?

Ans. The Dual Pressure Control have both high and low temperatures listed. The high side part of the control is a safety control in cases of a dirty condenser, a bad fan motor, overcharged system and several other reasons that could occur. In those cases, you would want the control to switch the unit off in the case of high discharge Pressure that can damage the valves or head gaskets of the compressor.

The other side, or Low-Pressure side of this control, also serves as a safety device that shuts the compressor off in the case of a loss of refrigerant. You always want the control to do this without letting the system run into a vacuum. The reason for this is that in a vacuum air will be sucked in introducing moisture into the system. Moisture is the #1 enemy of a refrigeration system. A good rule of thumb to remember is to set this control off at 2 lbs. of pressure and on at 30lbs of pressure. This setting will also work fine for a pump down off cycle.

Q. What is the oil safety control?

Ans. The Oil Safety Control is the least understood control in the system because it normally has three wires, and everyone panics. The two wires on the marked terminals L and M are only Control Circuit wires meaning they are usually the same wires that go to or from the low, high pressure control. They just series through the oil control again on terminals L & M. This also powers one side of the oil switch. The third wire is the one no one seems to understand what to do with. This wire is either on the 120V terminal or On the 230V terminal depending on the control voltage of the unit. The tricky part is that this wire MUST BE wired to the LOAD side of the contactor with the wires that feed the compressor (230V only) or through an Auxiliary Contact or Relay because it must shut off power to the Oil Control when the compressor is not running.

REFRIGERANTS

Refrigerant classifications are important to know as well as latent heats and freezing points for R22, 404a, 717 and 718.

FIGURE 3. Refrigerant safety classification from **ASHRAE** Standard 34.

	lower toxicity	higher toxicity	
higher flammability	A3	B3	LFL ≤ 0.10 kg/m² or heat of combustion ≥19 000kj/kg
lower flammability	A2 / A2L*	B2 / B2L*	LFL ≤ 0.10 kg/m² and heat of combustion ≥19 000kj/kg
no flame propagation	A1	B1	no LFL based on modified ASTM E681-85 test
	no identified toxicity at concentrations ≤400 ppm	evidence of toxicity below 400 ppm (based on data for TLV-TWA or consistent indices)	

*A2L and B2L are lower flammability refrigerants with a maximum burning velocity of < 10 cm/s.

There is no way you will be able to memorize the temperature/pressure chart (T/P). However, it will be well worth the time to spend a few minutes learning the ranges for the refrigerants and knowing which numbers correspond with the refrigerant market name. R718 is water. Water is actually a refrigerant. Water possesses properties that if used at the right pressure will operate no different than any other refrigerant. Just at a different temperature and pressure.

TEMPERATURE/PRESSURE CHART											
Pressure [PSIG/Inch Hg] REFRIGERANT			TEMP	TEMP	Pressure [PSIG] REFRIGERANT			TEMP	TEMP	Pressure REFRIG	
R-744	R-717	R-22			R-744	R-717	R-22			R-744	R-7
67.7	•	•	-24	-31.1	184.9	1.7	7.9	44	6.7	585.7	64
71.7	•	•	-20	-28.9	200.3	3.5	10.2	48	8.9	620.2	70
75.8	-	-	-16	-26.7	216.6	5.6	12.6	50	10.0	638.2	74
80.1	18.7"	11.9"	-12	-24.4	233.8	7.8	15.2	55	12.8	685.4	83
84.5	17.8"	10.9"	-8	-22.2	251.9	10.2	17.9	60	15.6	732.7	92
89	17.0"	9.8"	-4	-20.0	271	12.8	20.9	65	18.3	-	102
93.7	16.2"	8.6"	0	-17.8	291.1	15.6	24	70	21.1	-	113
98.6	15.3"	7.4"	4	-15.6	312.1	18.7	27.4	75	23.9	-	125
103.6	14.4"	6.1"	8	-13.3	334.2	22	31	80	26.7	-	13
108.8	13.3"	4.8"	12	-11.1	357.4	25.5	34.8	85	29.4	-	151
114.1	12.2"	3.4"	16	-8.9	381.7	29.3	38.8	90	32.2	-	165
119.6	11.1"	2.0"	20	-6.7	407.1	33.4	43.1	95	35.0	-	181
125.3	10.0"	0.4"	24	-4.4	433.7	37.7	47.6	100	37.8	-	197
131.2	8.8"	0.6	28	-2.2	461.6	42.4	52.4	105	40.6	-	214
143.5	6.1"	2.2	32	0.0	490.6	47.4	57.5	110	43.3	-	232
156.5	3.2"	3.9	36	2.2	521	52.7	62.9	115	46.1	-	251
170.3	0.0	5.9	40	4.4	552.6	58.4	68.6	120	48.9	-	271

R717 is NH3 or Ammonia. Yes, similar or the same ammonia that was used for domestic and commercial cleaning. However, refrigeration ammonia is at a higher concentration.

Operator safety is a requirement when working with ammonia. Gloves and masks for ventilation are a requirement. Observe the following minimum guidelines when handling refrigerants:

1. Ventilate the room if you suspect a leak
2. Keep the system at or below its design operating pressure
3. Do not mix refrigerants

4. 4. Do not allow flames near a system suspected of leaking--can create toxic gas

5. 5. Wear goggles and gloves, especially when charging or discharging a system

6. 6. Do not fill a service cylinder more than 80% full with refrigerant, as the fluid needs room to expand if temperature increases. A full cylinder is a bomb waiting to explode.

7. 7. Store cylinders in a cool place.

8. 8. Do not refill disposable cylinders

9. 9. Do not fill cylinders with anything except what is marked on the label.

10. 10. Do not sniff refrigerants--some can kill you.

11. 11. Do not depend on your nose to detect a leak--use an approved method. Halide torches, soap bubbles or electronic leak detectors are used depending on the type of refrigerant used.

Liquid refrigerant can quickly freeze the skin. Wash it off immediately with water and treat frostbite if needed.

CONCLUSION

Your studies beyond this point should be focused repetitive and focused on Memorandum or Outline of Topics for the licensure examination you are qualified to take. This book covers the basics of the components and accessories typically found on basic refrigeration systems.

Many refrigeration plants operate differently, and you will be expected to operate those plants in the manner recommended by the Plant Director and in accordance with their Standard Operating Procedures. Nevertheless, you are now equipped to have an intelligent conversation about the precepts and concepts of Basic Refrigeration.

Studying a chapter, a day and repeating the process will have you ready to take any basic refrigeration exam with confidence.

References

ASHRAE 1997 Fundamentals Handbook

Basic Refrigeration: Principles, Practices and Operation, Guy R. King

Refrigeration Licenses Unlimited, Clayton Carrico

www.ingramcontent.com/pod-product-compliance
Lightning Source LLC
Chambersburg PA
CBHW021456210526
45463CB00002B/796